IBALI ELAMANANI

THE NUMBER STORY

SMALL BOOK ONE

ENGLISH - XHOSA

Numbers Teach Children
Their Number Names

written and illustrated by

MISS ANNA

Early Reader Edition of *The Number Story 1*
Bronze Medal Winner, 2016 Wishing Shelf Book Award

Library of Congress Control Number: 2018902040

Names: Miss Anna, author.
Title: Number story : numbers teach children their number names / Miss Anna.
Description: Portland, OR: Lumpy Publishing, 2018.
Identifiers: ISBN 978-1-949320-13-8 | LCCN 2018902040
Summary: The pictures and rhymes present stories which introduce numbers 0-10.
Subjects: LCSH Numeration—English—Xhosa--Pictorial works--Juvenile literature. | BISAC JUVENILE NONFICTION /
Languages: English—Xhosa
Classification: LCC QA141.3 .M57 2018 | DDC 513—dc23

Publisher: Lumpy Publishing
Website: www.missannabooks.com
Email: missanna@missannabooks.com

Paperback: ISBN 978-1-949320-13-8
Printed in the U.S.A.   1 3 5 7 9 10 8 6 4 2

Uyafuna ukufunda amagama awa manani?

It is very easy and a lot of fun!

Kulula futhi kumnandi!

Say-along our little jingle

Cula nathi elibali ela manani!

starting from Number One!

Sizo qala kwi nani loku qala!

1

ONE   looks like my one finger.

INYE

Ingathi ifana nomnwe wam.

ONE!

INYE!

2

TWO trails a tail.

ZIMBINI

Utsala usmsila.

A TAIL! UMSILA!

3

THREE has bumps.

NTATHU

Inendawo ezijikayo.

BUMPY! IJIKA!

4

FOUR carries a sail.

EZINE

Uphata umkhumbi.

4
A SAIL!
NGOMKHUMBI!

5

FIVE   is a racing track.

EZINTLANU

Yindlela yokugijima.

VROOM
VRUUUM!
1

6

SIX   curves like a snail.

EZINTANDATHU

Ijikeleza njenge sneyli.

A SNAIL! ISNEYLI!

7

BE CAREFUL! IT'S SHARP!
LUMNKA! IBUKHALI!

8

EIGHT   is rollercoaster rails.

EZISIBHOZO

Ngu loliwe owe rola Kosta.

YEY!
YIPPEE!

NINE is a bubble on a stick.

EZITHOBA

Ibhabhul ekwi swazi.

A BUBBLE!

IBHABHULZ!

10

TEN   is an eye of a whale.

EZILISHUMI

Le linye ilihlo ele wheyle.

HELLO! MOLO!

And
Kunye

0

ZERO   is an empty pail.

UNOTHI

Li pheyle elingenanto.

IT'S EMPTY!
ALINANTO!

Thank you for playing with us today.

We had a lot of fun too!

Siyabulela ba nidlale nathi lamhlanje.

Seso nwaba nathi.

We are your Number friends,
Zero to Ten,
Who will be here for you~
Singama nani aba ngaba hlobo bakho
kunothi side siyofika kwezilishumi.

Sizo hlala silapha ngenxa yakho!

Bye-bye now!
See you again soon!
Bhayi bhayi kengoku!
Ndizokubona kungekudala!